POINTS OF V

Should We Keep ANIMALS in Zoos?

By Nick Christopher

Published in 2018 by
KidHaven Publishing, an Imprint of Greenhaven Publishing, LLC
353 3rd Avenue
Suite 255
New York, NY 10010

Designer: Seth Hughes
Editor: Katie Kawa

Photo credits: Cover © istockphoto.com/DragonImages; p. 5 (bottom) gans33/Shutterstock.com; p. 5 (top) Svietlieisha Olena/Shutterstock.com; p. 7 Guildhall Library & Art Gallery/Heritage Images/Getty Images; p. 9 belizar/Shutterstock.com; p. 11 © istockphoto.com/nymphae_fairy; p. 13 kojihirano/Shutterstock.com; p. 15 © istockphoto.com/andresr; p. 17 Image Source/Vetta/ Getty Images; p. 18 John Sommers II/Getty Images; p. 19 Tim Clayton/Corbis Documentary/ Getty Images; p. 21 (notepad) ESB Professional/Shutterstock.com; p. 21 (markers) Kucher Serhii/ Shutterstock.com; p. 21 (photo frame) FARBAI/Thinkstock; p. 21 (elephants) red-feniks/ Shutterstock.com; p. 21 (polar bear and giraffe) Purestock/Thinkstock; p. 21 (donkey) Joseph C. Salonis/Shutterstock.com.

Cataloging-in-Publication Data

Names: Christopher, Nick.
Title: Should we keep animals in zoos? / Nick Christopher.
Description: New York : KidHaven Publishing, 2018. | Series: Points of view | Includes index.
Identifiers: ISBN 9781534523258 (pbk.) | 9781534523272 (library bound) | ISBN 9781534523265 (6 pack) | ISBN 9781534523289 (ebook)
Subjects: LCSH: Captive wild animals–Juvenile literature. | Zoo animals–Juvenile literature. | Animal welfare–Juvenile literature.
Classification: LCC QL76.C47 2018 | DDC 590.74–dc23

Printed in the United States of America

CPSIA compliance information: Batch #BS17KL: For further information contact Greenhaven Publishing LLC, New York, New York at 1-844-317-7404.

Please visit our website, www.greenhavenpublishing.com. For a free color catalog of all our high-quality books, call toll free 1-844-317-7404 or fax 1-844-317-7405.

CONTENTS

Do Zoos HELP OR HURT?

Zoos are popular places in communities around the world. They're places where people can learn about wild animals by seeing them up close. Some people, though, believe zoos aren't good for animals. They think wild animals should only live in the wild.

Should animals be kept in zoos? Before you answer this question for yourself, it's helpful to understand why people on both sides of this **debate** feel the way they do. Seeing things from many points of view makes it easier to form an educated, or informed, opinion about important issues.

Know the Facts!

There are more than 10,000 zoos around the world today.

Some people see zoos as places that keep animals safe and healthy. Others see zoos as places that keep animals from being free.

The Rise of ANIMAL RIGHTS

Wild animals have been kept in **captivity** for thousands of years. Before zoos, rich people kept menageries, or collections of wild animals, to show off their wealth and power. Then, the first modern zoos were built in Europe in the 1700s.

Over time, people began speaking out about the practice of keeping wild animals in zoos. The rise of the animal rights movement and groups such as People for the **Ethical** Treatment of Animals (PETA) have led many to question whether or not zoos are ethical.

Know the Facts!

The word "zoo" is short for "zoological park" or "zoological garden."

The word "zoo" was first used to describe London's zoological gardens in the 1800s.

Caring for ANIMALS

Zoos provide care for animals that they wouldn't receive in the wild. Animals are fed a healthy **diet** in zoos. In the wild, they might not be able to get enough food or the right food because of problems such as **habitat destruction**. Also, zoos have veterinarians who provide medical care for hurt or sick animals that would often be left to die in the wild.

In some cases, zoos help animals live longer. A 2016 study showed that most **mammals** live longer in zoos than they do in the wild.

Know the Facts!

Zoos help smaller animals that are prey for larger animals in the wild. They live longer because they aren't hunted for food in zoos.

Most zookeepers love the animals they work with and give them proper food and medical care. They also provide programs called enrichment that keep animals learning and active as they would be in the wild.

Mental Health CONCERNS

Although many mammals live longer in zoos, some mammals live longer in the wild. Elephants live nearly twice as long in the wild as they do in captivity. Scientists believe the causes for this include elephants in zoos becoming overweight and **stressed**.

The mental health of zoo animals is an area of growing concern. A number of zoo animals suffer from **depression** and **anxiety**. Their mental health suffers when they're forced to live in small spaces. Also, zoo animals go through stress when they're moved to new zoos and taken away from their family or pack.

Know the Facts!

"Zoochosis" is the word commonly used for the strange actions and mental problems that an animal displays because it's in captivity.

Zoo animals experience stress from people staring at them, talking loudly around them, and taking their picture all day.

Saving SPECIES

Zoos play an important part in saving endangered species, which are groups of animals that are in danger of becoming extinct, or dying out. **Breeding** programs in zoos around the world aim to increase the populations of endangered species. In most cases, the plan is to let those animals go back into the wild.

The Association of Zoos and Aquariums (AZA) is the leader in caring for endangered species. Its Saving Animals from Extinction (SAFE) program is working to help certain species stay alive through **conservation** plans and lifesaving care in its 230 **accredited** zoos.

Know the Facts!

AZA-accredited zoos meet the highest standards for animal care and conservation efforts.

The California condor, shown here, was saved from extinction thanks in part to the efforts of people working at the San Diego Zoo. The population of this species rose from 22 to more than 400 in 30 years!

Can't GO BACK

It may seem like a good thing that zoos **release** animals back into the wild, but there are often problems with this practice. Life in a zoo is very different from life in an animal's natural habitat. Zoo animals don't learn hunting skills since they're fed by zookeepers. They also don't need to stay away from predators in captivity, so they don't know what to do when faced with a predator in the wild.

Zoo animals also don't learn to fear people. This leads to them being hit by cars or shot by hunters in the wild.

Know the Facts!

A study done by British scientists showed that most carnivores, or meat-eating animals, born in captivity die when people try to release them back into the wild.

Animals that live in zoos don't learn the skills they would in the wild. This can be deadly if an animal is released back into the wild.

Teaching TOOLS

One of the most important things zoos are supposed to do is educate people about animals. When children visit zoos, it often sparks a lifelong love for animals. They get to see wild animals they would never see otherwise in a safe setting. This helps them form a deeper connection with nature.

Zoos show children that they share Earth with wild animals and that they must help care for endangered species and other creatures. This gives them a better understanding of why conservation efforts are important. Zoos also teach the public new things about animal habitats.

Know the Facts!

According to the AZA, more than 51 million students visit zoos in North America each year.

Zoos give children a chance to see animals in person instead of in a book or on a screen.

The Wrong LESSONS

How much do zoos actually teach people about animals? Some studies have shown that it's not as much as you'd think. In fact, a 2014 study showed that out of 3,000 children who visited a zoo, only 1,000 of them learned something new.

Groups such as PETA believe the biggest thing children learn at zoos is that it's fine to keep animals in cages instead of in the wild. Some people also worry that zoos make the public care less about conservation because people believe zoos are taking care of the problem of endangered animals.

Know the Facts!

In 2016, a gorilla named Harambe was killed at the Cincinnati Zoo after it grabbed a boy who fell into its habitat. This led many people to question if zoos were safe enough for both animals and visitors.

It's not easy for children to learn how animals act in the wild when they visit a zoo because the animals aren't in their natural habitats.

Safety or FREEDOM?

In recent years, people who run zoos have worked hard to build habitats that are more like the places animals live in the wild. However, these man-made habitats still aren't the same as the natural world. Animals are often safer in zoos, but is safety more important than freedom?

People who support zoos believe these places save endangered animals and educate people. Others believe zoos do more harm than good. After learning the facts about both sides of this debate, what do you think? Should we keep animals in zoos?

Know the Facts!

Some zoos clip the wings of birds so they can't fly away. This allows them to live outside, but it also keeps visitors from seeing what they're truly like in the wild.

Should we keep ANIMALS in ZOOS?

YES

- Zoos provide medical care for animals they wouldn't get in the wild.
- Zookeepers feed animals a healthy diet.
- Zoos try to teach people about animals and the environment.
- Zoos save endangered animals.
- Zoos help people learn to love animals.

NO

- Zoo animals' mental health can suffer from being in captivity.
- Zoo animals don't learn to hunt or to avoid being hunted, which makes it hard for them to live in the wild.
- People don't learn much from zoos.
- Zoo animals aren't free and live in unnatural habitats.
- Zoos teach people it's okay to keep animals in cages.

These are just some of the reasons people feel strongly in favor of and against zoos. Making a list can help you form your own point of view about important issues.

GLOSSARY

accredited: Officially said to have met a certain standard.

anxiety: Fear or nervousness.

breed: To produce young.

captivity: The state of being kept in a place and unable to escape.

conservation: The careful management of the natural world.

debate: An argument or discussion about an issue, generally between two sides.

depression: A mental disorder marked by sadness and a lack of interest in doing anything.

diet: The food and drink an animal takes in.

ethical: Meeting accepted standards for what is right and wrong.

habitat destruction: The loss of an area where plants and animals live because the land can no longer support those living things.

mammal: Any warm-blooded animal that has babies that drink milk and a body covered with hair or fur.

release: To set free.

stressed: Made to feel mental pressure.

For More INFORMATION

WEBSITES

Association of Zoos & Aquariums
www.aza.org
The AZA's official website offers a closer look at its SAFE program and other conservation programs, as well as a list of accredited zoos and aquariums you can visit.

PETA Kids
www.petakids.com
PETA's official website for young people includes information on issues related to animal rights, including problems with zoos.

BOOKS

O'Connell, Caitlin, and T.C. Rodwell. *Bridge to the Wild: Behind the Scenes at the Zoo*. Boston, MA: Houghton Mifflin Harcourt, 2016.

Parker, Vic. *Let's Think About Animal Rights*. Chicago, IL: Heinemann Library, 2015.

Shea, Therese. *Discovering STEM at the Zoo*. New York, NY: PowerKids Press, 2016.

Publisher's note to educators and parents: Our editors have carefully reviewed these websites to ensure that they are suitable for students. Many websites change frequently, however, and we cannot guarantee that a site's future contents will continue to meet our high standards of quality and educational value. Be advised that students should be closely supervised whenever they access the Internet.

INDEX